Speed,time, and distance

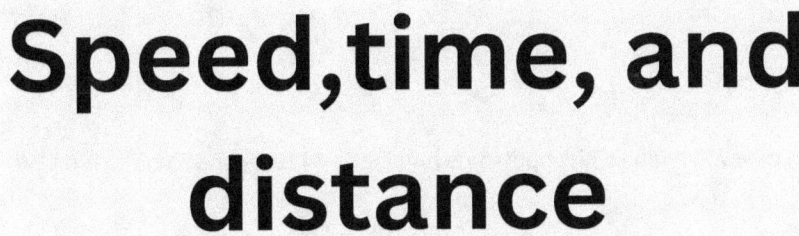

Wordproblems
Primary kids

How long will it take a hippopotamus running at 44 feet per second to run 176 yards?

_____

_____

A conveyor can move materials a distance of 371 meters in 53 seconds. How fast does the conveyor move?

_____

_____

A flock of migratory birds is flying at 35 mph. How far will they fly in 10 hours?

_____

_____

How long will it take Lynette to drive from Roseland to Quicksburg driving at an average speed of 93 kilometers per hour if the distance is 372 kilometers?

_____

_____

A train travels 834 kilometers in 6 hours. What is its average speed?

_____

_____

How far will an airplane fly in 3 hours at a cruising speed of 520 miles per hour?

_____

_____

Answer Key

How long will it take a hippopotamus running at 44 feet per second to run 176 yards?

176 yards × 3 = 528 feet

528 feet ÷ 44 feet per second = 12 seconds

A conveyor can move materials a distance of 371 meters in 53 seconds. How fast does the conveyor move?

371 meters ÷ 53 seconds = 7 meters per second

A flock of migratory birds is flying at 35 mph. How far will they fly in 10 hours?

35 mph × 10 hours = 350 miles

How long will it take Lynette to drive from Roseland to Quicksburg driving at an average speed of 93 kilometers per hour if the distance is 372 kilometers?

372 kilometers ÷ 93 kph = 4 hours

A train travels 834 kilometers in 6 hours. What is its average speed?

834 kilometers ÷ 6 hours = 139 kph

How far will an airplane fly in 3 hours at a cruising speed of 520 miles per hour?

520 mph × 3 hours = 1560 miles

How long will it take a flock of migratory birds to fly 195 miles if they fly at 39 mph?

How long will it take an airplane flying at an average cruising speed of 450 miles per hour to fly 900 miles?

How far can a conveyor moving at 369 meters per minute move materials in 18 minutes?

An airplane flies 4200 kilometers in 5 hours at cruising speed. What is its average cruising speed?

A train travels 207 miles in 3 hours. What is its average speed?

How far will a bus travel in 2 hours at an average speed of 85 kilometers per hour?

Answer Key

A flock of migratory birds is flying at 36 mph. How far will they fly in 9 hours?

36 mph × 9 hours = 324 miles

A black bear can run at a speed of 37 feet per second. How many yards can it run in 21 seconds?

37 feet per second × 21 seconds = 777 feet

777 feet ÷ 3 = 259 yards

An airplane flies 1980 kilometers in 3 hours at cruising speed. What is its average cruising speed?

1980 kilometers ÷ 3 hours = 660 kph

Pete is driving from Bevington to Hawthorne, a distance of 658 kilometers. If he takes 7 hours to get to Hawthorne, what is his average speed?

658 kilometers ÷ 7 hours = 94 kph

How long will it take an airplane flying at an average cruising speed of 180 kilometers per hour to fly 540 kilometers?

540 kilometers ÷ 180 kph = 3 hours

How long will it take a coveyor moving at 2 meters per second to move materials 88 meters?

88 meters ÷ 2 meters per second = 44 seconds

How long will it take an airplane flying at an average cruising speed of 260 kilometers per hour to fly 1560 kilometers?

_____

A train travels 148 kilometers in 2 hours. What is its average speed?

_____

Jordan drove from Ashfield to Redrock in 4 hours. If he averaged 70 miles per hour, how far is it from Ashfield to Redrock?

_____

A flock of migratory birds can fly 38 miles in 2 hours. How fast do they fly?

_____

How far can a conveyor moving at 388 meters per minute move materials in 4 minutes?

_____

How long will it take a grizzly bear running at 44 feet per second to run 2948 yards?

_____

Answer Key

How long will it take an airplane flying at an average cruising speed of 260 kilometers per hour to fly 1560 kilometers?

1560 kilometers ÷ 260 kph = 6 hours

A train travels 148 kilometers in 2 hours. What is its average speed?

148 kilometers ÷ 2 hours = 74 kph

Jordan drove from Ashfield to Redrock in 4 hours. If he averaged 70 miles per hour, how far is it from Ashfield to Redrock?

70 mph × 4 hours = 280 miles

A flock of migratory birds can fly 38 miles in 2 hours. How fast do they fly?

38 miles ÷ 2 hours = 19 mph

How far can a conveyor moving at 388 meters per minute move materials in 4 minutes?

388 meters per minute × 4 minutes = 1552 meters

How long will it take a grizzly bear running at 44 feet per second to run 2948 yards?

2948 yards × 3 = 8844 feet
8844 feet ÷ 44 feet per second = 201 seconds

How far can a conveyor moving at 1125 feet per minute move materials in 15 minutes?

A train travels 80 miles in 4 hours. What is its average speed?

A cheetah can run at a speed of 103 feet per second. How many yards can it run in 12 seconds?

A flock of migratory birds can fly 208 miles in 8 hours. How fast do they fly?

How long will it take an airplane flying at an average cruising speed of 620 kilometers per hour to fly 1240 kilometers?

How long will it take an airplane flying at an average cruising speed of 200 miles per hour to fly 400 miles?

Answer Key

How far can a conveyor moving at 1125 feet per minute move materials in 15 minutes?

1125 feet per minute × 15 minutes = 16875 feet

A train travels 80 miles in 4 hours. What is its average speed?

80 miles ÷ 4 hours = 20 mph

A cheetah can run at a speed of 103 feet per second. How many yards can it run in 12 seconds?

103 feet per second × 12 seconds = 1236 feet
1236 feet ÷ 3 = 412 yards

A flock of migratory birds can fly 208 miles in 8 hours. How fast do they fly?

208 miles ÷ 8 hours = 26 mph

How long will it take an airplane flying at an average cruising speed of 620 kilometers per hour to fly 1240 kilometers?

1240 kilometers ÷ 620 kph = 2 hours

How long will it take an airplane flying at an average cruising speed of 200 miles per hour to fly 400 miles?

400 miles ÷ 200 mph = 2 hours

A flock of migratory birds is flying at 36 mph. How far will they fly in 9 hours?

A black bear can run at a speed of 37 feet per second. How many yards can it run in 21 seconds?

An airplane flies 1980 kilometers in 3 hours at cruising speed. What is its average cruising speed?

Pete is driving from Bevington to Hawthorne, a distance of 658 kilometers. If he takes 7 hours to get to Hawthorne, what is his average speed?

How long will it take an airplane flying at an average cruising speed of 180 kilometers per hour to fly 540 kilometers?

How long will it take a coveyor moving at 2 meters per second to move materials 88 meters?

Answer Key

A flock of migratory birds is flying at 36 mph. How far will they fly in 9 hours?

36 mph × 9 hours = 324 miles

A black bear can run at a speed of 37 feet per second. How many yards can it run in 21 seconds?

37 feet per second × 21 seconds = 777 feet

777 feet ÷ 3 = 259 yards

An airplane flies 1980 kilometers in 3 hours at cruising speed. What is its average cruising speed?

1980 kilometers ÷ 3 hours = 660 kph

Pete is driving from Bevington to Hawthorne, a distance of 658 kilometers. If he takes 7 hours to get to Hawthorne, what is his average speed?

658 kilometers ÷ 7 hours = 94 kph

How long will it take an airplane flying at an average cruising speed of 180 kilometers per hour to fly 540 kilometers?

540 kilometers ÷ 180 kph = 3 hours

How long will it take a coveyor moving at 2 meters per second to move materials 88 meters?

88 meters ÷ 2 meters per second = 44 seconds

An airplane flies 960 miles in 4 hours at cruising speed. What is its average cruising speed?

How far will a train travel in 6 hours if it is moving at an average speed of 172 kilometers per hour?

How long will it take Missie to drive from Lawson to Tremont driving at an average speed of 111 kilometers per hour if the distance is 222 kilometers?

A flock of migratory birds is flying at 37 mph. How far will they fly in 8 hours?

A bus traveling on a freeway travels 384 miles in 6 hours. What is its average speed?

How long will it take a coveyor moving at 345 meters per minute to move materials 3795 meters?

Answer Key

An airplane flies 960 miles in 4 hours at cruising speed. What is its average cruising speed?

960 miles ÷ 4 hours = 240 mph

How far will a train travel in 6 hours if it is moving at an average speed of 172 kilometers per hour?

172 kph × 6 hours = 1032 kilometers

How long will it take Missie to drive from Lawson to Tremont driving at an average speed of 111 kilometers per hour if the distance is 222 kilometers?

222 kilometers ÷ 111 kph = 2 hours

A flock of migratory birds is flying at 37 mph. How far will they fly in 8 hours?

37 mph × 8 hours = 296 miles

A bus traveling on a freeway travels 384 miles in 6 hours. What is its average speed?

384 miles ÷ 6 hours = 64 mph

How long will it take a coveyor moving at 345 meters per minute to move materials 3795 meters?

3795 meters ÷ 345 meters per minute = 11 minutes

How far will a bus travel in 3 hours at an average speed of 93 kilometers per hour?

How long will it take a flock of migratory birds to fly 315 miles if they fly at 35 mph?

How far can a conveyor moving at 11 feet per second move materials in 34 seconds?

An airplane flies 1800 miles in 5 hours at cruising speed. What is its average cruising speed?

A train travels 366 kilometers in 6 hours. What is its average speed?

How long will it take a hippopotamus running at 44 feet per second to run 220 yards?

Answer Key

How far will a bus travel in 3 hours at an average speed of 93 kilometers per hour?

93 kph × 3 hours = 279 kilometers

How long will it take a flock of migratory birds to fly 315 miles if they fly at 35 mph?

315 miles ÷ 35 mph = 9 hours

How far can a conveyor moving at 11 feet per second move materials in 34 seconds?

11 feet per second × 34 seconds = 374 feet

An airplane flies 1800 miles in 5 hours at cruising speed. What is its average cruising speed?

1800 miles ÷ 5 hours = 360 mph

A train travels 366 kilometers in 6 hours. What is its average speed?

366 kilometers ÷ 6 hours = 61 kph

How long will it take a hippopotamus running at 44 feet per second to run 220 yards?

220 yards × 3 = 660 feet
660 feet ÷ 44 feet per second = 15 seconds

A bus traveling on a freeway travels 420 kilometers in 5 hours. What is its average speed?

_____

_____

An airplane flies 2120 kilometers in 4 hours at cruising speed. What is its average cruising speed?

_____

_____

How long will it take Christy to drive from Whitetop to Bear Lake driving at an average speed of 65 kilometers per hour if the distance is 195 kilometers?

_____

_____

How far can a conveyor moving at 262 feet per minute move materials in 11 minutes?

_____

_____

How far will an airplane fly in 5 hours at a cruising speed of 180 miles per hour?

_____

_____

How long will it take a train moving at an average speed of 46 miles per hour to travel 322 miles?

_____

_____

Answer Key

A bus traveling on a freeway travels 420 kilometers in 5 hours. What is its average speed?

420 kilometers ÷ 5 hours = 84 kph

An airplane flies 2120 kilometers in 4 hours at cruising speed. What is its average cruising speed?

2120 kilometers ÷ 4 hours = 530 kph

How long will it take Christy to drive from Whitetop to Bear Lake driving at an average speed of 65 kilometers per hour if the distance is 195 kilometers?

195 kilometers ÷ 65 kph = 3 hours

How far can a conveyor moving at 262 feet per minute move materials in 11 minutes?

262 feet per minute × 11 minutes = 2882 feet

How far will an airplane fly in 5 hours at a cruising speed of 180 miles per hour?

180 mph × 5 hours = 900 miles

How long will it take a train moving at an average speed of 46 miles per hour to travel 322 miles?

322 miles ÷ 46 mph = 7 hours

A conveyor can move materials a distance of 3256 feet in 11 minutes. How fast does the conveyor move?

How long will it take an airplane flying at an average cruising speed of 860 kilometers per hour to fly 4300 kilometers?

An airplane flies 2240 miles in 4 hours at cruising speed. What is its average cruising speed?

A moose can run at a speed of 51 feet per second. How many yards can it run in 24 seconds?

How far will a bus travel in 7 hours at an average speed of 80 kilometers per hour?

How long will it take Marlene to drive from Satsop to Dexter driving at an average speed of 73 miles per hour if the distance is 365 miles?

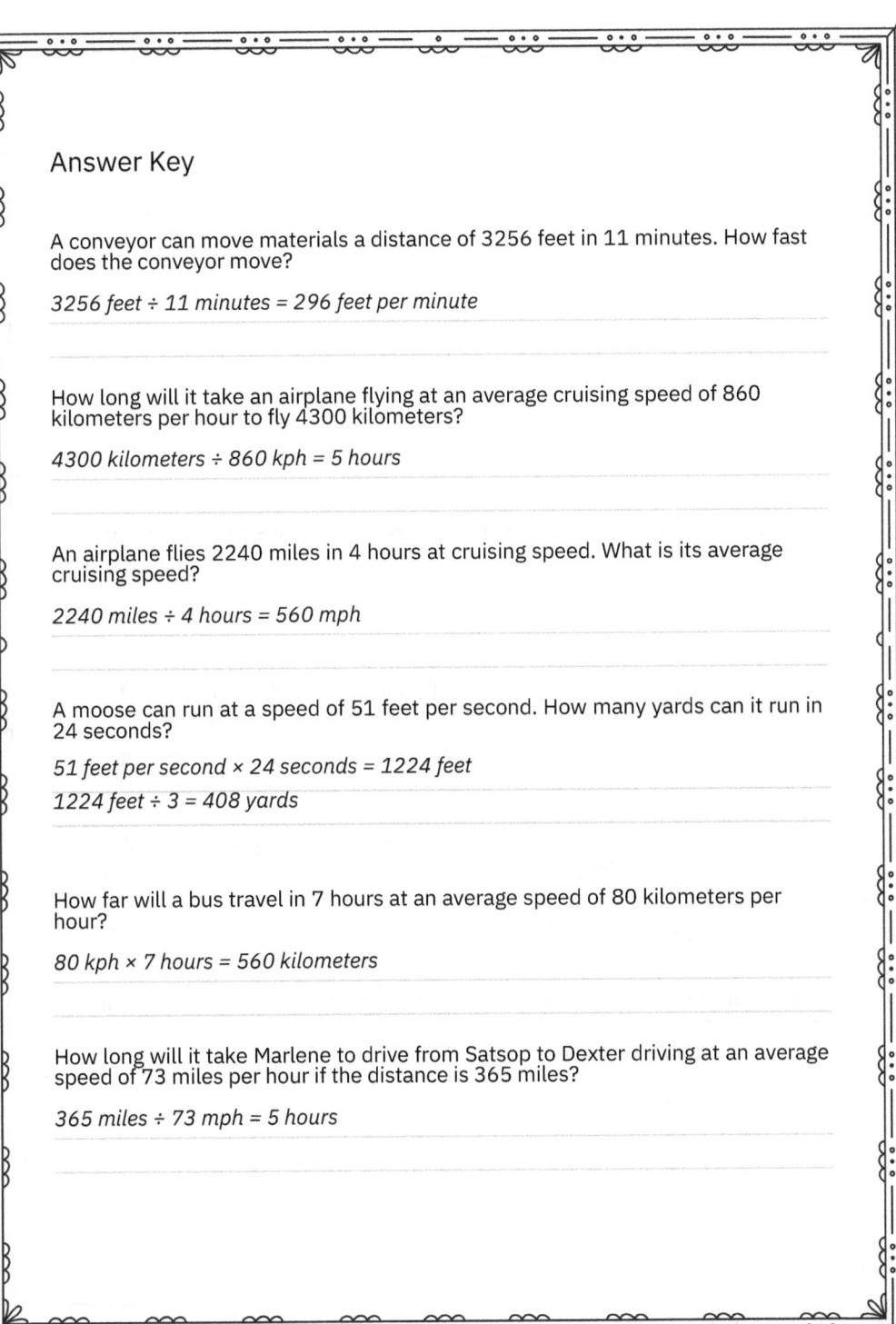

Answer Key

A conveyor can move materials a distance of 3256 feet in 11 minutes. How fast does the conveyor move?

3256 feet ÷ 11 minutes = 296 feet per minute

How long will it take an airplane flying at an average cruising speed of 860 kilometers per hour to fly 4300 kilometers?

4300 kilometers ÷ 860 kph = 5 hours

An airplane flies 2240 miles in 4 hours at cruising speed. What is its average cruising speed?

2240 miles ÷ 4 hours = 560 mph

A moose can run at a speed of 51 feet per second. How many yards can it run in 24 seconds?

51 feet per second × 24 seconds = 1224 feet

1224 feet ÷ 3 = 408 yards

How far will a bus travel in 7 hours at an average speed of 80 kilometers per hour?

80 kph × 7 hours = 560 kilometers

How long will it take Marlene to drive from Satsop to Dexter driving at an average speed of 73 miles per hour if the distance is 365 miles?

365 miles ÷ 73 mph = 5 hours

How far will an airplane fly in 6 hours at a cruising speed of 810 kilometers per hour?

_____

A flock of migratory birds can fly 189 miles in 7 hours. How fast do they fly?

_____

How long will it take a train moving at an average speed of 82 miles per hour to travel 246 miles?

_____

Joanna is driving from Colburn to Wesson, a distance of 240 miles. If she takes 5 hours to get to Wesson, what is her average speed?

_____

How long will it take an airplane flying at an average cruising speed of 390 miles per hour to fly 780 miles?

_____

A wild boar can run at a speed of 44 feet per second. How many yards can it run in 21 seconds?

_____

Answer Key

How far will an airplane fly in 6 hours at a cruising speed of 810 kilometers per hour?

810 kph × 6 hours = 4860 kilometers

A flock of migratory birds can fly 189 miles in 7 hours. How fast do they fly?

189 miles ÷ 7 hours = 27 mph

How long will it take a train moving at an average speed of 82 miles per hour to travel 246 miles?

246 miles ÷ 82 mph = 3 hours

Joanna is driving from Colburn to Wesson, a distance of 240 miles. If she takes 5 hours to get to Wesson, what is her average speed?

240 miles ÷ 5 hours = 48 mph

How long will it take an airplane flying at an average cruising speed of 390 miles per hour to fly 780 miles?

780 miles ÷ 390 mph = 2 hours

A wild boar can run at a speed of 44 feet per second. How many yards can it run in 21 seconds?

44 feet per second × 21 seconds = 924 feet
924 feet ÷ 3 = 308 yards

How long will it take an airplane flying at an average cruising speed of 290 kilometers per hour to fly 870 kilometers?

How far will a bus travel in 7 hours at an average speed of 119 kilometers per hour?

How long will it take a train moving at an average speed of 29 kilometers per hour to travel 145 kilometers?

How far will a bullet flying 5000 feet per second fly in 0.018 seconds?

A flock of migratory birds can fly 184 miles in 8 hours. How fast do they fly?

Shirley is driving from Stoddard to Belforest, a distance of 290 miles. If she takes 5 hours to get to Belforest, what is her average speed?

Answer Key

How long will it take an airplane flying at an average cruising speed of 290 kilometers per hour to fly 870 kilometers?

870 kilometers ÷ 290 kph = 3 hours

How far will a bus travel in 7 hours at an average speed of 119 kilometers per hour?

119 kph × 7 hours = 833 kilometers

How long will it take a train moving at an average speed of 29 kilometers per hour to travel 145 kilometers?

145 kilometers ÷ 29 kph = 5 hours

How far will a bullet flying 5000 feet per second fly in 0.018 seconds?

5000 feet per second ÷ 3 = 1666.667 yards per second
1666.667 yards per second × 0.018 seconds = 30 yards

A flock of migratory birds can fly 184 miles in 8 hours. How fast do they fly?

184 miles ÷ 8 hours = 23 mph

Shirley is driving from Stoddard to Belforest, a distance of 290 miles. If she takes 5 hours to get to Belforest, what is her average speed?

290 miles ÷ 5 hours = 58 mph

An airplane flies 1080 kilometers in 2 hours at cruising speed. What is its average cruising speed?

How long will it take a polar bear running at 29 feet per second to run 696 yards?

A flock of migratory birds is flying at 21 mph. How far will they fly in 2 hours?

An airplane flies 4250 kilometers in 5 hours at cruising speed. What is its average cruising speed?

Evelyn is going for a walk. How long will it take her to walk 18.75 miles if her average speed is three and three quarters miles per hour?

How far will a bus travel in 7 hours at an average speed of 91 kilometers per hour?

Answer Key

An airplane flies 1080 kilometers in 2 hours at cruising speed. What is its average cruising speed?

1080 kilometers ÷ 2 hours = 540 kph

How long will it take a polar bear running at 29 feet per second to run 696 yards?

696 yards × 3 = 2088 feet
2088 feet ÷ 29 feet per second = 72 seconds

A flock of migratory birds is flying at 21 mph. How far will they fly in 2 hours?

21 mph × 2 hours = 42 miles

An airplane flies 4250 kilometers in 5 hours at cruising speed. What is its average cruising speed?

4250 kilometers ÷ 5 hours = 850 kph

Evelyn is going for a walk. How long will it take her to walk 18.75 miles if her average speed is three and three quarters miles per hour?

18.75 miles ÷ 3.75 miles per hour = 5 hours

How far will a bus travel in 7 hours at an average speed of 91 kilometers per hour?

91 kph × 7 hours = 637 kilometers